Oxford International Primary Geography

Workbook

2

Terry Jennings

OXFORD
UNIVERSITY PRESS

Great Clarendon Street, Oxford, OX2 6DP, United Kingdom

Oxford University Press is a department of the University of Oxford. It furthers the University's objective of excellence in research, scholarship, and education by publishing worldwide. Oxford is a registered trade mark of Oxford University Press in the UK and in certain other countries

© Terry Jennings 2015

The moral rights of the authors have been asserted

First published in 2015

All rights reserved. No part of this publication may be reproduced, stored in a retrieval system, or transmitted, in any form or by any means, without the prior permission in writing of Oxford University Press, or as expressly permitted by law, by licence or under terms agreed with the appropriate reprographics rights organization. Enquiries concerning reproduction outside the scope of the above should be sent to the Rights Department, Oxford University Press, at the address above.

You must not circulate this work in any other form and you must impose this same condition on any acquirer

British Library Cataloguing in Publication Data
Data available

978-019-831-010-5

16 15 14

Paper used in the production of this book is a natural, recyclable product made from wood grown in sustainable forests. The manufacturing process conforms to the environmental regulations of the country of origin.

Printed in Great Britain by Ashford Colour Press Ltd. Gosport

Acknowledgements

The publishers would like to thank the following for permissions to use their photographs:

Cover photo: Getty Images/Stephan Studd

Although we have made every effort to trace and contact all copyright holders before publication this has not been possible in all cases. If notified, the publisher will rectify any errors or omissions at the earliest opportunity.

Links to third party websites are provided by Oxford in good faith and for information only. Oxford disclaims any responsibility for the materials contained in any third party website referenced in this work.

Contents

YEAR 2

1 THE WIDER WORLD

The Earth	4
Continents	6
Countries	8
Oceans and seas	10
My environment	12
Recycling at home	14

2 AN ISLAND HOME

Islands	16
Bahrain, an island country	18
Island transport	20

3 GOING TO THE SEASIDE

At the seaside	22
Seaside cities, towns and villages	24
Seasides around the world	26
Food from the sea	28

4 HOW WE LEARN ABOUT THE WORLD

Making sense of the world	30
Helping our eyesight	32
Globes, maps and atlases	34

5 PASSPORT TO THE WORLD

Climate	36
The global supermarket	38
China	40
Living in Shanghai	42

Map of the World — 44

Glossary — 46

The Earth

The parts of the Earth

Label the parts of the Earth using the words from the word box.

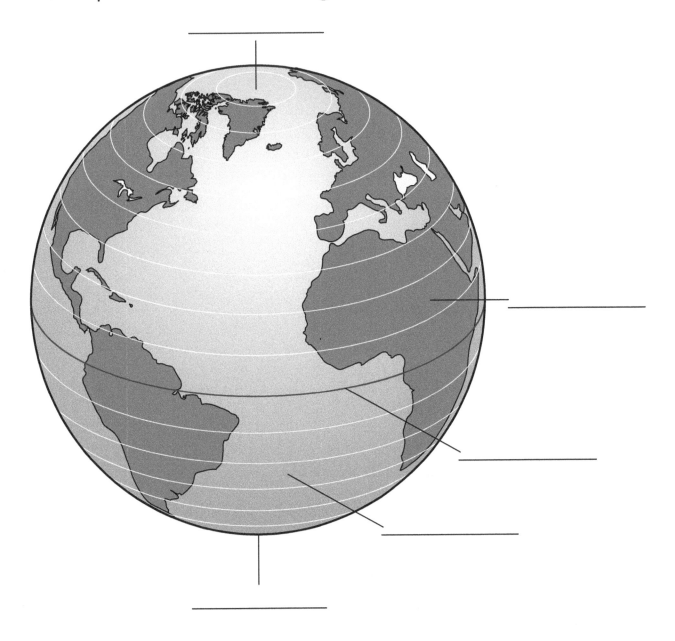

| North Pole | South Pole | Equator |
| Continent | Ocean | |

Why do we have night?

Fabian wanted to show his friends why we have night.
This is what he did.

What did Fabian use? _____

_____

What did he do? _____

_____

What did Fabian's friends see? _____

_____

Fabian's friends had to imagine that the globe was the Earth. What did they have to imagine the torch was? _____

_____

Continents

Continents

This is a map of the world.

Colour the land brown or green.

Colour the oceans and seas blue.

The seven large pieces of land are the continents.

Using the words from the word box, write the name of the continents on the map. Use an atlas to help you.

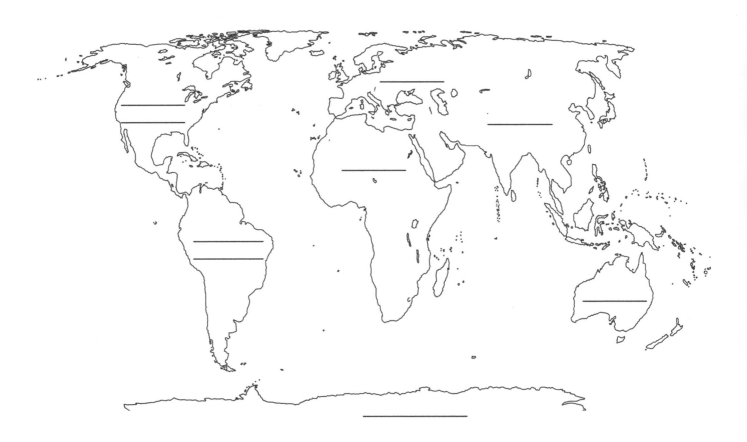

| Africa | Asia | Antarctica | Europe |
| North America | South America | Oceania | |

The seven continents

Here are pictures of six continents.

Colour them brown or green.

Label each continent. Use an atlas to help you.

Which continent is missing? _____

Countries

Kenya, an African country

There are over 50 countries in Africa.

One of these countries is Kenya.

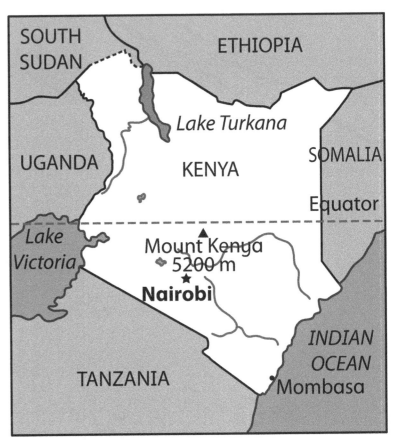

Look at the map of Kenya. Answer these questions:

What is the capital city of Kenya? _____

Which ocean lies to one side of Kenya? _____

What is the name of the imaginary line that runs through Kenya?

_____

What is the height of Mount Kenya? _____

Which two lakes are partly in Kenya? _____

Make your own country

Imagine you could make your own country.

What is the name of your country? _____

Is your country small or large? _____

Which continent is your country in? _____

What language is spoken there? _____

What is the landscape like? Is it hot and dry like a desert, does it have lots of mountains, or is it surrounded by sea? _____
_____

Draw your country's flag.

Draw a map or picture of your country.

Oceans and seas

The oceans

Look at the map of the world.

Label the five oceans, using the words from the word box.

Atlantic Ocean   **Arctic Ocean**   **Pacific Ocean**
Indian Ocean   **Southern Ocean**

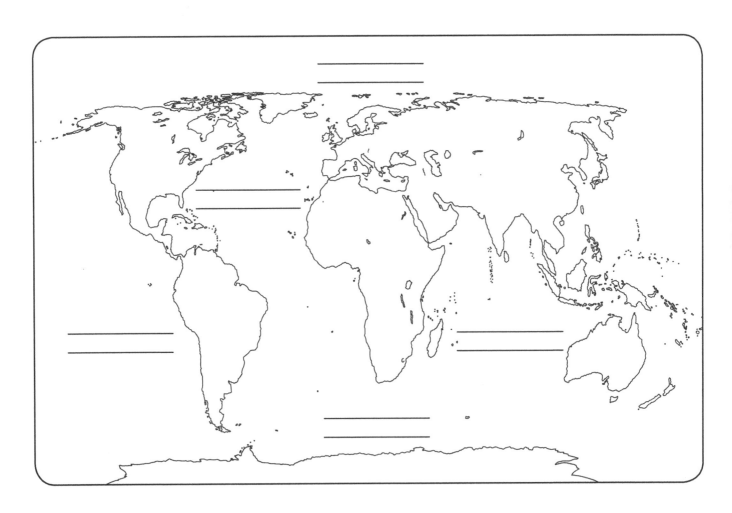

Oceans, continents and seas

Continents are huge areas of land.

Oceans are huge areas of salt water.

Seas are smaller areas of salt water.

Sort the words from the word box. Write the correct word in each column.

Oceania	Southern	Pacific	Mediterranean	Europe
Indian	South China	Red	Arabian	Africa
Dead	Asia	Atlantic	North America	Arctic

Continents	Oceans	Seas

The wider world

My environment

My environment

All the things around you make up your environment.

Some of these things are living.

Some of them are not living – or non-living.

In the picture below, colour all the things that are living.

Use a different colour for the things that are non-living. Make a key for your picture.

Draw yourself in the picture.

The environment and people

Some parts of our environment have been made by people.

Colour the pictures, then label them using the words from the word box.

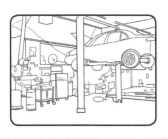

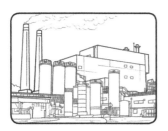

| garage | factory | house | playground | supermarket |
| road | shop | school | garden | |

Which of these things do you have in your environment? _____
_____

Recycling at home

Recycle it!

Some of the things we throw away can be recycled.

This means the materials they are made of can be used again.

We can recycle paper, glass, metals and many plastics.

Draw lines to match the objects with the correct recycling bin.

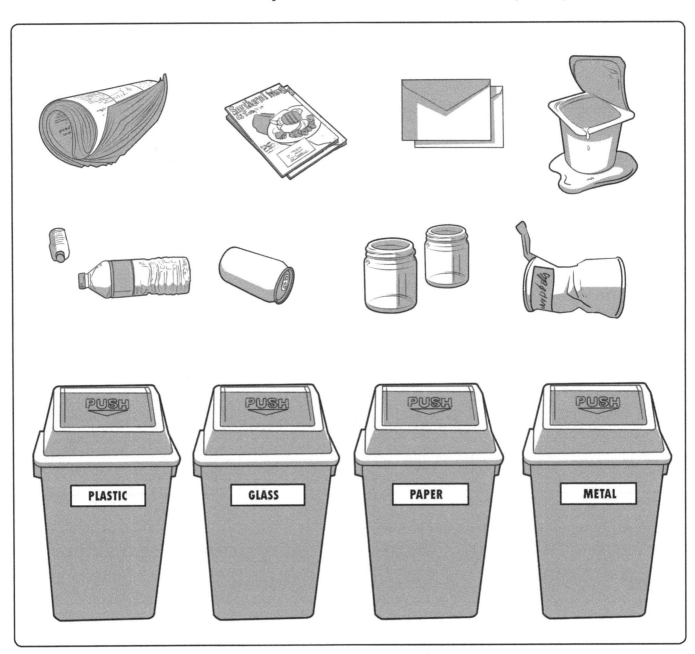

Reusing and recycling

Look at the items that are often found in dustbins.

Say what each item is made from.

Say how each item could be reused or recycled.

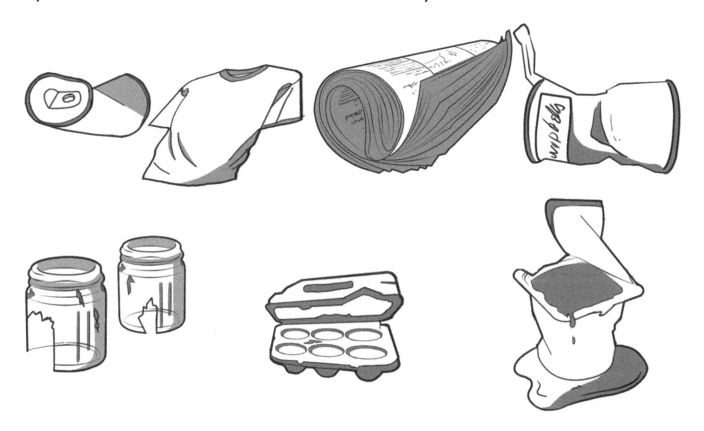

Item	What is it made from?	How could it be reused or recycled?

Islands

Islands

How many islands are there in this picture? _____

How do you know they are islands? _____

Colour the land in the picture green.

Colour the sea blue.

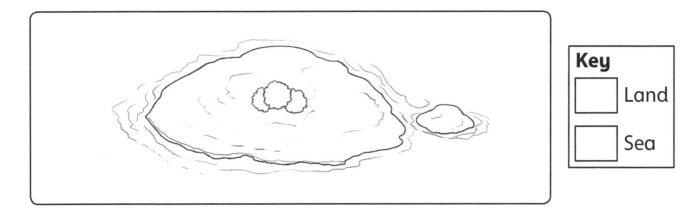

Key
☐ Land
☐ Sea

This is a map of the islands.

Colour the land on the map green.

Colour the sea blue.

Colour the key to the map.

Island life

Look in an atlas.

Choose an island where you would like to live.

Use books or the Internet to find out about your island.

How is the island different from your local area? Write eight things that are different.

	My local area	My island
1		
2		
3		
4		
5		
6		
7		
8		

Bahrain, an island country

Bahrain, an island country

This is a map of the island country of Bahrain.

Some of the islands are too small to be shown on this map.

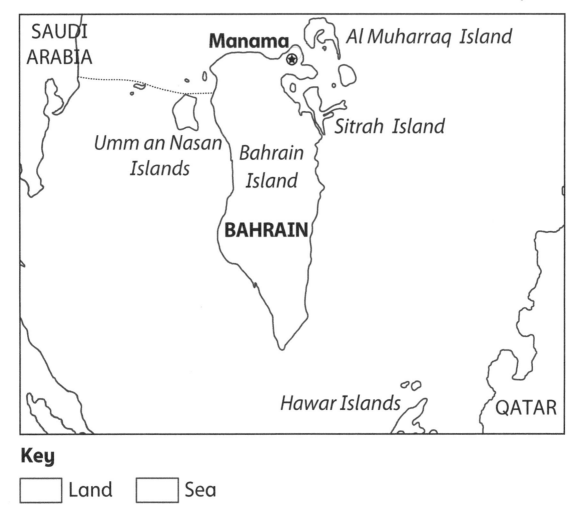

Key

☐ Land   ☐ Sea

Colour the islands green.

Colour the sea blue.

Colour the key to your map.

Which is the largest island on the map? _____

How many islands can you see on your map? _____

What is the name of the capital city of Bahrain? _____

Comparing islands

Choose an island country. It could be a small island country such as Singapore or Malta, or a large island country such as Greenland or Australia.

Write the island country at the top of the third column. Fill in the information for the island and then compare the island with Bahrain.

Name of island country	Bahrain	
Number of islands	33	
Name of largest island	Bahrain	
Capital city	Manama	
Population	About 1.3 million	
Currency (money)	Bahrain dinar	
Landscape	Mainly low desert	
Main exports	Oil and natural gas	
Flag	White Red	

An island home

Island transport

Island transport

This is a map of the island of Mull, a small island off the coast of Scotland.

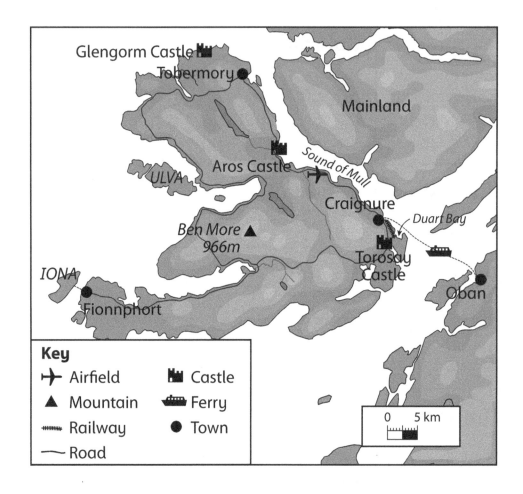

Look carefully at the map.

The key shows what the symbols mean.

What types of transport are there on Mull?

Write them in a list or draw pictures.

Transport from Bahrain

This is a map of Bahrain and part of the mainland of Saudi Arabia and Qatar.

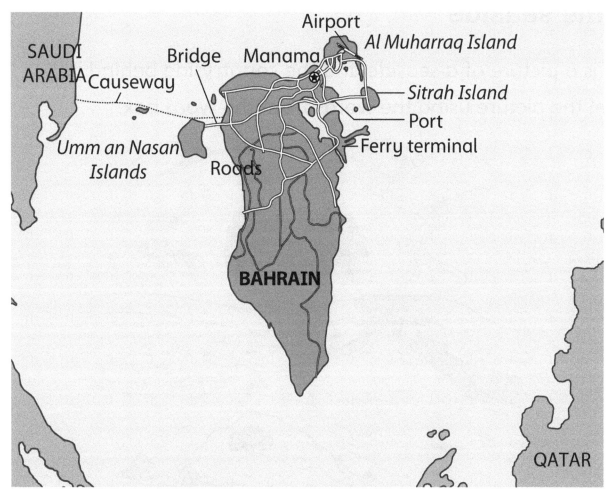

Look at the map carefully.

What is the quickest and easiest way to travel from Bahrain to the places below?

Would you travel by car or bus, ferry or plane? Write your answer beside each place name.

Saudi Arabia _____

Al Muharraq Island _____

Sitrah Island _____

Qatar _____

Paris, France _____

At the seaside

At the seaside

This is a picture of a seaside and the countryside behind it.
Label the picture using the words from the word box.

sea	beach	cliff	island	hill
river	bridge	forest	road	farm

How is sea water different from water in a river? _____

_____

What are some of the different things that beaches are made of?

_____

On the beach

Some things have been washed up on a beach.

How many of them can you name?

Sort the things into the two groups given in the list.

Natural things were not made by people.

Rubbish is things people throw away because they do not want them.

Natural things	Rubbish

Choose one of the pieces of rubbish on the beach.

How do you think it got there? _____

Why is it harmful? _____

_____

Seaside cities, towns and villages

A seaside holiday

Some people go on holiday to the seaside to enjoy themselves.

Some people work at the seaside.

Look at the picture. Find three people who are working.

Find three people who are enjoying themselves on holiday.

Draw or write about what the people are doing.

People on holiday	People working

A postcard from the seaside

Imagine you are on holiday by the seaside.

This postcard shows a picture of the area where you are staying.

On the back of the postcard write the name and address of a friend.

Describe what you have been doing and what you have seen.

Seasides around the world

Seaside words and pictures

Look at the words in the word box. Each one is the name of something that might be found in, on or near the sea.

Choose four of the words. In the spaces below, draw a picture and write one sentence about each word.

| whale | pebble | fish | ship | seaweed |
| sand | octopus | dolphin | crab | seashell |

A seaside holiday brochure

Choose a seaside resort in another country.

Make a page for a holiday brochure describing it.

COME TO _____

Picture or photograph of your chosen seaside resort

The weather is _____

It is a great place because _____

_____

There is lots to do. You can _____

_____

It is best to travel here by _____

Food from the sea

A fish's journey

Colour these pictures. Use the words from the word box to complete the sentences.

The fishing boat pulls a big _____ through the water.

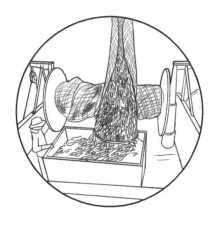

The fish are emptied from the net and kept _____ until the boat gets back to _____.

A lorry takes the fish to the nearest city.

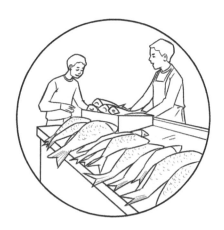

The fish are sold in a _____.

What other types of shop sell fish? _____

| port | market | frozen | net |

Animals from the sea

Colour the animals that live in the sea.

Going to the seaside

Making sense of the world

My senses

You have five main senses.

What are they?

Write the names of the senses on the picture.

Use your senses to compare two different places.

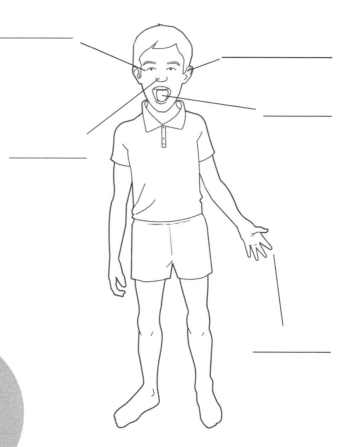

Do NOT use your sense of taste.

Sense	Place:	Place:
What I can see		
What I can hear		
What I can smell		
What I can touch		

A local survey

Carry out a survey of a busy place near your home or school.

Where did you carry out your survey? _____

Date and time of your survey: _____

Draw pictures of, or write words about, what you could see, hear, smell and feel.

What I could see	What I could hear

What I could smell	What I could feel

How many people did you see? _____

Helping our eyesight

Helping our eyesight

These are some of the things that help us to learn more about our world.

 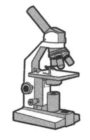 

Telescope	Microscope	Magnifying glass

Satellite	Camera

Which of these things help to make far away objects seem nearer?

_____

Which of these things make small objects seem larger?

_____

Which of these things takes photographs of the Earth from space?

_____

Which of these things lets us take pictures of the world around us?

_____

How observant are you?

Here are two copies of the same map.

There are **ten** differences between the two maps.

Draw a circle, on the second map, around each thing that is different.

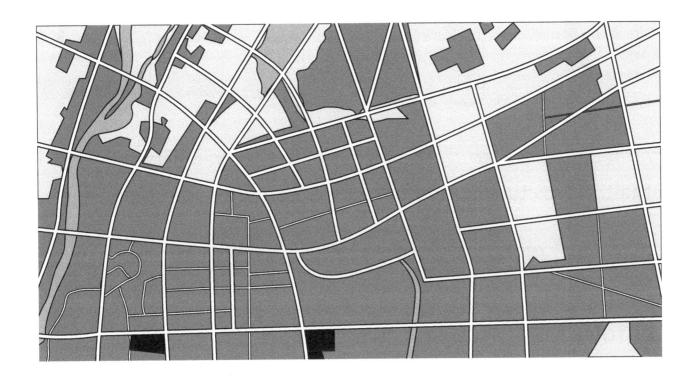

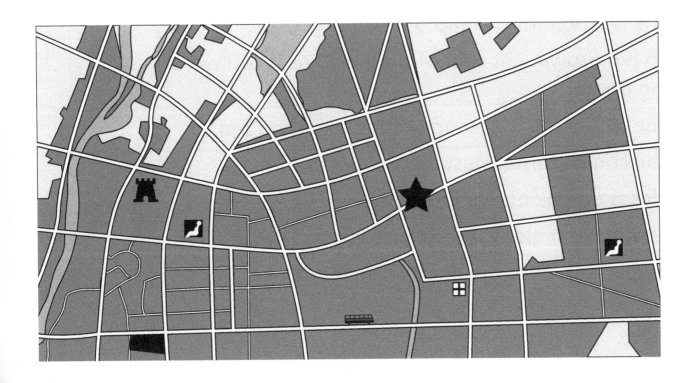

Globes, maps and atlases

Globes, maps and atlases

  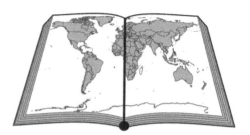

_____   _____   _____

Label the three pictures, using the words from the word box.

> map   atlas   globe

True or false?

Look at these sentences. For each one, put ✓ in the box if you think it is right, or ✗ if you think it is wrong.

A globe is a ball with a map of the world on it. ☐

A map is a drawing of part of the world as if you were looking up at it. ☐

A globe shows the true shape of the land and sea. ☐

A globe can be turned, just like the Earth turns. ☐

An atlas is a book of maps. ☐

On a map and a globe, everything is bigger than it really is. ☐

The maps in an atlas are large-scale maps. ☐

Using maps

Look at these two maps.

They show the same place but have different scales.

 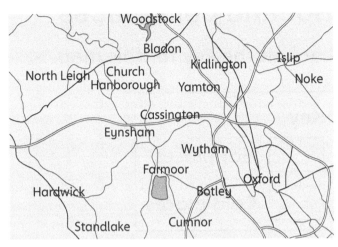

_____   _____

Label each map using one of the labels from the word box.

> **large-scale map**   **smaller-scale map**

Which map would you use if you were out walking? _____

Why? _____

_____

Which map would you use if you were driving a car? _____

Why? _____

_____

Climate

Hot and cold places

Look at the key for this map.

Key	
☐	Hot places
☐	Cold places

Colour the 'Hot places' box red.

Colour the 'Cold places' box blue.

Now colour the map of the world to show hot and cold places.

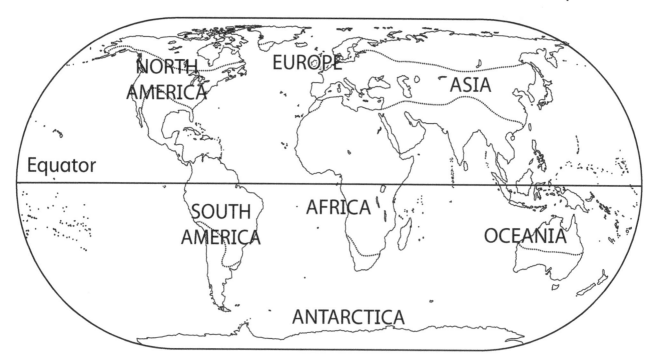

The young explorers

Imagine you and a friend are explorers.

You are going to a hot desert.

Your friend is going to the cold Arctic.

Which of these things will you pack?

Which of these things will your friend pack?

| Shorts | Trousers | Sun hat | Woolly hat |

| Flip-flops | Snow boots | Sunglasses | Snow goggles |

| Water bottle | T-shirt | Warm coat | Mittens |

Going to a hot desert	Going to the cold Arctic

The global supermarket

The global supermarket

The labels on cans and packets often tell us where our food has come from.

Work with a group of friends.

Look at as many different food containers as you can.

For each one write down the details like this:

Food	Country	Continent	Container
Peaches	South Africa	Africa	Can

If you have a computer, make a spreadsheet of your information using the headings shown above.

Look at all your results and answer these questions:

1 From which continent does most of our food come? _____

2 Which continent provides the least amount of food? _____

3 From which country does most of our food come? _____

4 List five foods that are produced in your country. _____

_____

Draw a block graph showing which food comes from each continent.

Food from around the world

Wissam and Sophie went to the supermarket.

They wrote down where some of the food came from.

They made a block graph of their results.

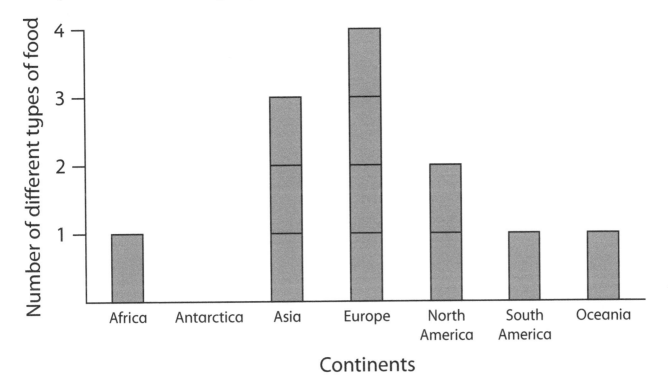

Answer these questions about Wissam and Sophie's graph.

How many types of food came from Asia?

_____

How many types of food came from Africa?

_____

From which continent did most food come?

_____

From which continent did no food come?

_____

China

Where is China?

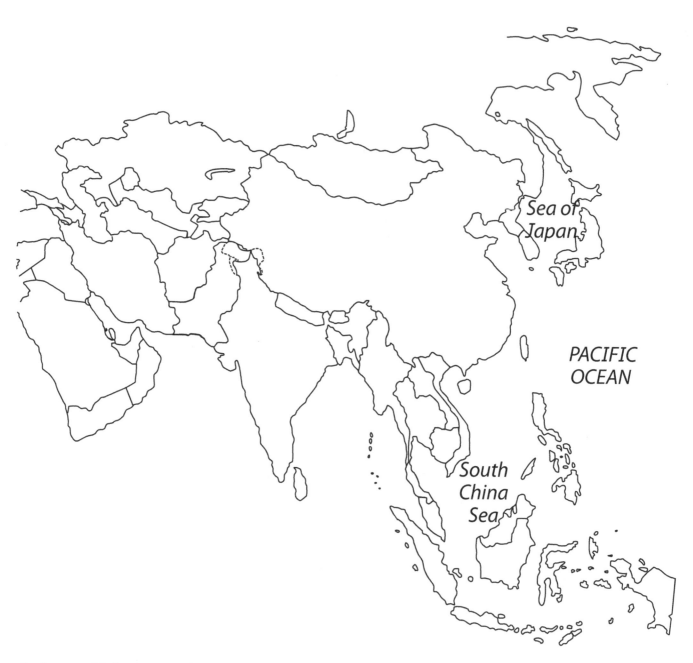

Colour China on the map.

Colour some of the other countries you know.

Write the names of the countries on your map. You could use an atlas to help you.

All about China

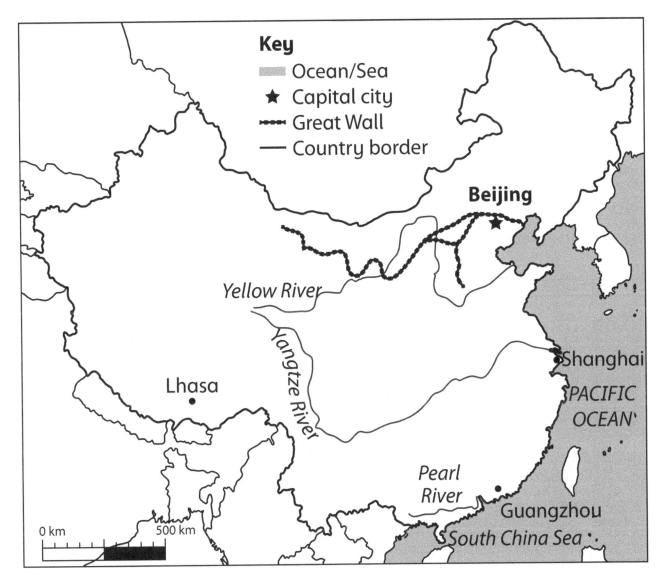

Answer these questions about China.

1 What is the capital of China? _____

2 Which ocean would you cross if you sailed away from Shanghai?
   _____

3 Which river does the Great Wall cross? _____

4 Which big river has a gemstone in its name? _____

5 Name six countries that border China. _____

Living in Shanghai

Living in China

Many people in China live in large cities like Shanghai.

Other people live in villages in the countryside.

What do you think you would see, hear and smell in a city and in the countryside, in China?

Shanghai	Chinese village
See	See
Hear	Hear
Smell	Smell

China fact file

Find out all you can about China.

Fill in the fact file below.

Continent	
Population	
Capital city	
Currency (money)	
Flag	
Languages spoken	
Other big cities	
Famous sights and interesting facts	

Map of the World

44

Glossary

Artificial satellite — A machine sent into space to help us communicate with one another or to collect information about the Earth.

Atlas — A book of maps.

Beach — The strip of sand, shingle, mud or rock where an ocean, sea or lake meets the land.

Border — A line that marks the edge of a country.

Capital — The most important city in a country.

Cargo — A load of goods carried by a lorry, train, aircraft or ship.

Causeway — A raised road or walkway over water.

Cliff — A steep wall of rock, especially on the coast.

Climate — The typical weather of a place over a whole year.

Coast — The seashore and the land close to it.

Coastline — The line on a map marking where the sea meets the land.

Continent — One of the seven big pieces of land in the world.

Coral — A hard substance made from the shells of tiny sea animals.

Desert — A large area of land where few plants can grow because it is either too dry or too cold.

Equator — A line drawn on maps to show places half-way between the North Pole and the South Pole.

Ferry — A ship used for carrying people or

	things across a river or narrow sea.	**Port**	A harbour or a town or city with a harbour.
Globe	A ball with a map of the whole world on it.	**Recycle**	To treat waste material so that it can be used again.
Government	The group of people who are in charge of a country.	**Resort**	A place where people go for their holidays.
Harbour	An inlet of the sea which gives ships a place to unload or shelter from bad weather.	**Sand**	The tiny grains of rock that you find on beaches and in deserts.
Iceberg	A very large block of ice floating in the sea.	**Seaside**	A place, such as a village, town or city, by the sea.
Lens	A curved piece of glass or plastic used to make things look larger or smaller.	**Sense**	The ability to see, hear, touch, taste or smell.
		Sphere	A globe; the shape of a ball.
Mainland	The main part of a country, not the islands around it.	**Tide**	The rising and falling of the level of the sea, which happens twice a day. This is caused mainly by the Moon's gravity pulling the water on the Earth.
Map	A drawing of part or all of the Earth's surface as if you were looking down on it.		
Mountain	A very high part of the Earth's surface.		

Valley	A line of low land between hills or mountains.	**Wave**	A moving ridge of water on the sea, formed by the wind blowing over the water.
Volcano	A mountain with a hole at the top through which hot, molten rock comes out from deep inside the Earth.	**Weather**	The rain, wind, snow and sunshine, for example, at a particular time or place.